CAPTAIN INVINCIBLE
and the
SPACE SHAPES

by Stuart J. Murphy ✶ illustrated by Rémy Simard

To Sam Manzi—whose mother, Phoebe Yeh,
is an invincible MathStart captain
—S.M.

For the Godmar boys
—R.S.

The publisher and author would like to thank teachers Patricia Chase, Phyllis Goldman, and Patrick
Hopfensperger for their help in making the math in MathStart just right for kids.

HarperCollins®, ▰®, and MathStart® are registered trademarks of HarperCollins Publishers. For
more information about the MathStart series, write to HarperCollins Children's Books,
10 East 53rd Street, New York, NY 10022, or visit our website at www.mathstartbooks.com.

Bugs incorporated in the MathStart series design were painted by Jon Buller.

Library of Congress Cataloging-in-Publication Data
Murphy, Stuart.
  Captain Invincible and the space shapes / by Stuart Murphy ; illustrated by Rémy Simard.
    p.   cm.—(MathStart)
  "Level 2."
  Summary: While piloting his spaceship through the skies, Captain Invincible encounters three-
dimensional shapes, including cubes, cylinders, and pyramids.
  ISBN 0-06-028022-0 — ISBN 0-06-028023-9 (lib. bdg.) — ISBN 0-06-446731-7 (pbk.)
    1. Geometry—Juvenile literature. [1. Geometry. 2. Shape.] I. Simard, Rémy, ill. II. Title.
III. Series.
QA445.5 .M857 2001   00-039609
516—dc21

           Typography by Elynn Cohen 13 14 15 16 SCP 20 19 18 17 16 ❖ First Edition

CAPTAIN INVINCIBLE

and the
SPACE SHAPES

Watch out!
If the rocks hit us,
we'll never make it home.

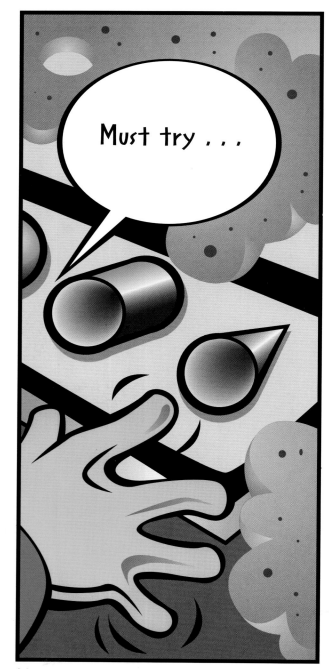

We're going to be okay, Comet. The cone will pull the gas in through its base, shaped like a circle—

and send out clean air through its tip.

Ah . . . good as new.

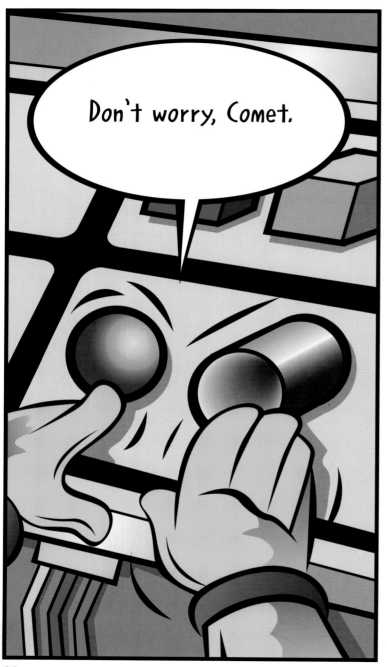

24

25

"What's going on?" asked Dad.
"You woke everyone up."
"Hey," said Brad. "Look at your spaceship.
It's all broken."
"Go to bed, Sam," Mom said, yawning.
"You can clean up this mess tomorrow."

In *Captain Invincible and the Space Shapes* the math concept is identifying three-dimensional shapes. Recognizing and classifying shapes like the cube, pyramid, and cylinder lays a foundation for understanding geometry.

If you would like to have more fun with the math concepts presented in *Captain Invincible and the Space Shapes*, here are a few suggestions:

- Before reading the story, talk to the child about a trip on a spaceship and the meaning of the word *invincible*. Make a copy of the instrument panel (page 7). As you read the story with your child, have him or her look on the instrument panel and locate the shape that Captain Invincible has pushed.

- Reread the story and have the child look for the different three-dimensional shapes throughout the art.

- Ask the child, "How is the square different from the other shapes in the same row on the instrument panel?" Then discuss with the child the similarities and differences of all the shapes in the square row. Continue by discussing the circle row.

- At the end of the story, Sam dreams of being Captain Stupendous, King of the Seas. Have the child make up a story about Captain Stupendous and how the captain might use the six shapes.

Following are some activities that will help you extend the concepts presented in *Captain Invincible and the Space Shapes* into a child's everyday life:

Scavenger Shape Hunt: Make a chart similar to the one shown below. Have the child look around the house and find objects that are like the shapes in the story. Record the name of the object under the name of the three-dimensional shape.

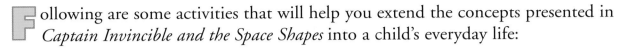

Cube	Rectangular Prism	Pyramid	Sphere	Cylinder	Cone

Spaceship: Have the child create his or her own spaceship using the six shapes found in the story. You can make shapes out of construction paper or use shapes found around the house to construct the spaceship.

Space Riddles: Make up riddles about the attributes of the various space shapes, for example, "I have six faces and they are all the same. Who am I?" Let the child try to guess the riddles and encourage him or her to create some for others to guess.

The following stories include concepts similar to those that are presented in *Captain Invincible and the Space Shapes*:

- THE GREEDY TRIANGLE by Marilyn Burns

- WHEN A LINE BENDS . . . A SHAPE BEGINS by Rhonda Gowler Greene

- CUBES, CONES, CYLINDERS, AND SPHERES by Tana Hoban